Colour Atlas of Trees and Woody Plants

Colour Atlas of Trees and Woody Plants

Bryan G. Bowes

CRC Press
Taylor & Francis Group
Boca Raton London New York

CRC Press is an imprint of the
Taylor & Francis Group, an **informa** business

First edition published 2020
by CRC Press
6000 Broken Sound Parkway NW, Suite 300, Boca Raton, FL 33487-2742

and by CRC Press
2 Park Square, Milton Park, Abingdon, Oxon, OX14 4RN

ISBN: 978-0-8153-9250-7 (hbk)
ISBN: 978-0-367-47398-3 (pbk)
ISBN: 978-1-351-11079-2 (ebk)

Typeset in Times
by Deanta Global Publishing Services, Chennai, India

Dedication

———————

To my late wife Ruth and my wife Diane; my children Tanya and Adrian Bowes and my stepchildren James and Rowan Eisner.

Contents

Preface: Trees and Why We Love Them

It is only since retiring as an academic botanist that I have fully appreciated how important trees are to me. I then became a Recorder for the Tree Register, inspecting and measuring large, unrecorded trees in the UK. In my early years, trees were perceived by me as an enhancement to the landscape and rather taken for granted. Since then I have realised the immense importance of trees to us all and how essential they are for preserving environmental diversity.

In 1869, John Muir wrote in *My First Summer in the Sierra* (Houghton Mifflin, Boston 1911):

> After the excursion to Mount Hoffman I had seen a complete cross section of the Sierra forest, and I find that *Abies magnifica* is the most symmetrical tree of all the noble coniferous company. The cones are grand affairs, superb in form, size, and colour, cylindrical, stand erect on the upper branches like casks, and … covered with fine down which has a silvery lustre in the sunshine, and their brilliance is augmented by beads of transparent balsam which seems to have been poured over each cone … If possible, the inside of the cone is more beautiful than the outside; the scales, bracts, and seed wings are tinted with the loveliest rosy purple with a bright lustrous iridescence … When the cones are ripe the scales and bracts fall off, setting the seed free to fly to their predestined places, while the dead spike-like axes are left on the branches to mark the positions of the vanished cones …

A reverence and respect for trees, such as is exhibited in Muir's scientific but beautiful prose, is deeply embedded in the human psyche. A sacred oak (*Quercus*) was at the centre of the Greek oracle at Dodona, founded in about 1800 BCE, whilst Artemis, the goddess of woodlands, and Apollo, her twin brother, were born under a date palm. For the Pehuenche Amerindians of Chile, the seeds of the Monkey Puzzle (*Araucaria araucana*) remain a staple part of their diet and the tree is regarded as holy. In India, Buddha is believed to have attained enlightenment whilst sitting under the bodhi tree (*Ficus religiosa*). Ginkgo (*Ginkgo biloba*) is almost extinct in the wild in China but has survived as a sacred tree in temple gardens.

However, for the last century or more, such respect for trees has generally been forsaken. Even today, unsustainable logging and forest clearances continue at an increased rate, either illegally or with official connivance, as in present-day Brazil. Nevertheless, the rise of numerous environmental advocacy movements, such as Greenpeace, Friends of the Earth, Extinction Rebellion, and many other organisations, that protest the desecration of nature give us considerable, albeit last-minute, hope for the future.

Acknowledgements

The writing of this book has given me the opportunity to share my love of trees with a wider audience and show photographs of the many special trees which I have taken over my long life as a botanist. I would like to acknowledge the unstinting computer advice and other help given by Mark Partridge (formerly IT Adviser at University of Glasgow). I wish to thank Norman Tait (formerly Departmental Photographer, University of Glasgow) for his technical advice over many years of friendship. I also wish to warmly thank my daughter, Tanya, for her expert proof reading of this manuscript, my grandson Aidan for his internet wizardry, and my wife, Diane, for always being there and who shares my love of trees.

Author

Dr Bryan G. Bowes is a retired Senior Lecturer in Botany at the University of Glasgow (1964–89) and author/editor of various books on plant morphology, anatomy, and ultrastructure that were originally published by Manson Press, London, now part of CRC Publishing, Boca Raton, Florida. In addition, he is the author of various scientific articles covering important aspects of plant morphogenesis in which he has developed an especial interest in the course of his nearly 60-year scientific career. Since retirement, Dr Bowes has become a keen nature photographer and most of the pictures which illustrate the current book were taken by him during varied trips/excursions in the UK or in locations abroad.

1 Introduction

LAND PLANTS

All land plants contain chloroplasts, which colour their tissues, as seen in the newly expanding green leaves (Figure 1.1) of *Aesculus hippocastanum* (Horse Chestnut). Chloroplasts are the minute organelles – visible in detail under an electron microscope (Figure 1.2) – where photosynthesis occurs.

BRIEF SURVEY OF ANCIENT LAND PLANTS

A layer of chert, a rock rich in silica, is located at Rhynie in Aberdeenshire, Scotland. The rock's microscopic structure was first investigated by Kidston and Lang, working in Glasgow University between 1917 and 1921. Both they and later workers reported on the discovery of fossilised remnants of an early land flora which existed near Rhynie some 410 million years ago (mya). The plant remains were petrified *in situ* by ash from volcanoes which polluted local pools of water where, at their margins, various very small primitive land plants were then growing. The remains of this Rhynie flora (preserved in the rocky chert) contain randomly preserved fragments of some of the earliest known land plants. Figure 1.3 shows the polished surface of a chert block in which plant debris is clearly visible; in Figure 1.4 a section of the rock is viewed under a light microscope, whereas a longitudinal section of the spore-containing capsule of *Aglaophyton* – one of these primitive plants – is illustrated in Figure 1.5.

From these and other primitive land plants, a varied land flora evolved, and, by the later Carboniferous Period, a rich arboreal flora existed; numerous remains of this flora have been identified from locations throughout the world. One such site is illustrated in Figure 1.6, which shows a fossil-rich limestone quarry on the outskirts of Glasgow, Scotland. In Figure 1.7 a fragment of the limestone debris lying at the quarry base reveals the petrified remains of a *Lepidodendron* (*Stigmaria*) root, with numerous broken-off stubs of its rootlets visible on its surface, whereas others pass out into what was the original soil surface.

In the rich Carboniferous plant deposits located in Scotland, further important fossils have been discovered. Figure 1.8 shows the protective shelter built around Fossil Grove at Victoria Park, Glasgow. During the laying-out of the recreation grounds at Victoria Park in the 19th century, workmen exposed the petrified stumps (still lying *in situ*) of several large trees, which were then identified as *Lepidodendron* fossils. Due to their exceptional scientific importance, a protective shelter was erected around them and the surrounding rock was carefully chiselled away to reveal their internal stone casts. On the viewing platform of this shelter, a group of University undergraduates are studying several fossil casts of the tree remains, which lie exposed on the rock floor. In Figure 1.9 a more detailed view of one of these casts reveals how, at the original swampy ground level, the vertical tree trunk divided dichotomously (equally) into horizontal roots. They ran for several metres along the surface of the lagoon where this stand of trees was growing. Figure 1.10 is a diagrammatic reconstruction of a mature specimen of *Lepidodendron*. Such trees are

FIGURE 1.1

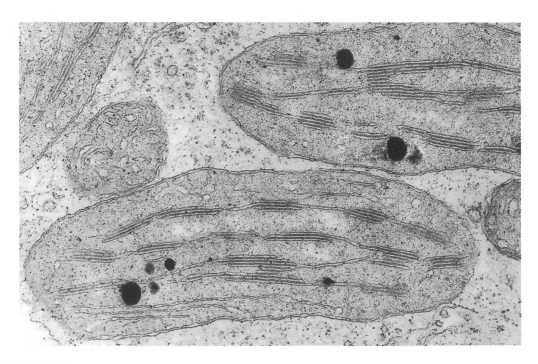

FIGURE 1.2

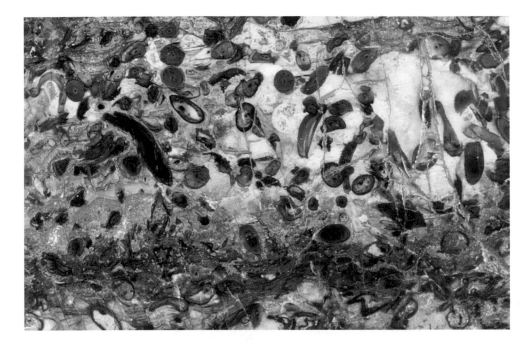

FIGURE 1.3

FIGURE 1.4

FIGURE 1.5

FIGURE 1.6

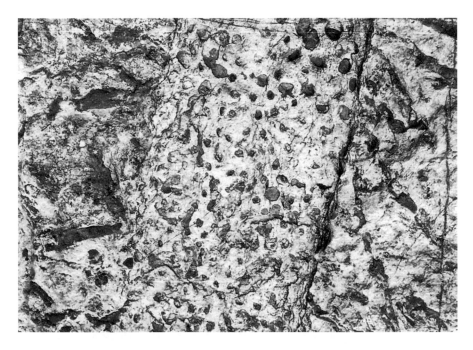

FIGURE 1.7

FIGURE 1.8

FIGURE 1.9

FIGURE 1.10

estimated to have reached some 30 m in height, with their closely crowded trunks some 1 m wide at the base. The tree trunk rarely branched, except at its tip, which bore branches bearing clusters of narrow leaves and reproductive cones. The trunk's layer of bark (composed of persistent photosynthetic leaf cushions on its surface; Figure 1.11) played an important role in mechanical support for the tree. Unfortunately, during the excavation of these fossils, their bark was destroyed by the chiselling away of the enclosing rock matrix. Within the trunk there was some central secondary wood composed of radially arranged tracheids (Figure 1.12).

By the Carboniferous Period, ferns bearing seeds were also present in the flora: Figure 1.13 shows a rock section with the stem of the seed fern *Lyginopteris oldhamia* revealing its well-developed secondary xylem cylinder, enclosing a prominent pith. Figure 1.14 shows a fossilised seed of *Lagenostoma ovoides*, which is surrounded by a thick carbonised seed coat.

By the later Triassic Period (some 200–250 mya), extensive tropical forests of large coniferous trees had developed, with some trees estimated to reach 60 m in height, with trunks up to 0.6 m wide. Figure 1.15 shows the trunk of *Araucarioxylon arizonicum* (revealed by the erosion of the surrounding rock) in the Petrified Forest of Northern Arizona, USA. The trunk is now lying horizontally and supported on a pillar of underlying, non-eroded rock. Such fossilised wood is often brightly coloured with red and yellow hues, caused by deposits on it of

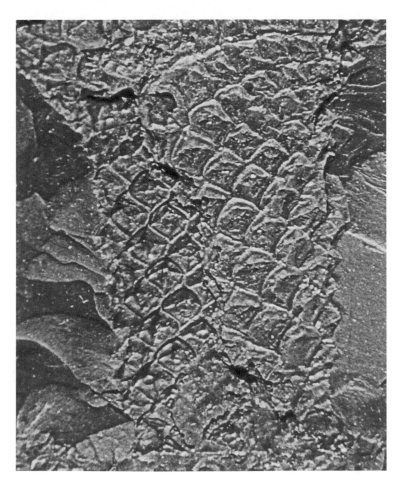

FIGURE 1.11

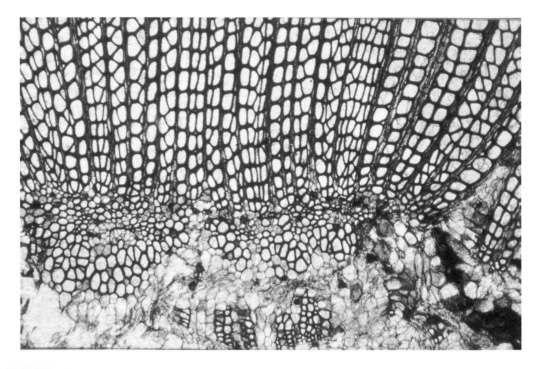

FIGURE 1.12

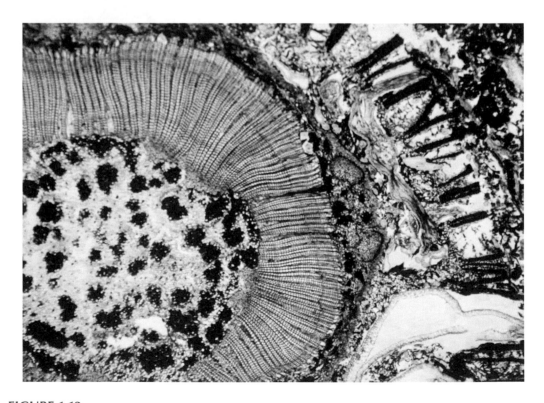

FIGURE 1.13

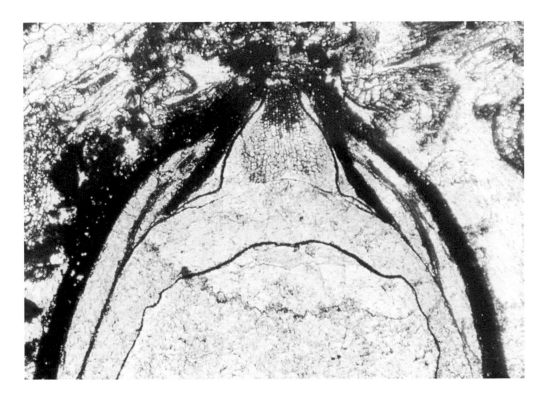

FIGURE 1.14

FIGURE 1.15

haematite and limonite particles. The genus *Araucarioxylon* probably comprises several different conifers, but the poor preservation of the wood makes their discrimination very difficult.

More recently formed plant fossil deposits contain both coniferous and flowering plant remains. Figure 1.16 illustrates the carbonised remains of a leaf from a fossil leaf bed (Figure 1.17) in the Lower Tertiary (about 60 mya). This mudstone deposit was discovered in 1851 by the Duke of Argyll, at Ardtun on the Isle of Mull, Scotland. The leaf shows the reticulate venation of a dicot and greatly resembles the present-day genus *Corylus* (Hazel). This mudstone also contains abundant remains of other leaves, apparently related to present-day trees such as *Platanus* (Plane), *Quercus* (Oak), and *Ginkgo* (Maidenhair tree).

FIGURE 1.16

FIGURE 1.17

BRIEF SURVEY OF PRESENT-DAY TREES AND WOODY PLANTS

TREE FERNS

There are some 300 species of arborescent ferns, principally in the genera *Cyathea* and *Dicksonia*, which mainly grow in tropical mountainous regions but are also found in temperate Australasia and elsewhere. Figure 1.18 shows a stand of mature specimens of the tree fern *Dicksonia antartica* growing in the mild climate experienced at Logan Botanic Garden, Port Logan, Scotland. In its native Australian habitat, *Dicksonia* sometimes occurs in expansive stands where it grows slowly and may attain some 15 m in height. The trunk is usually unbranched and densely enclosed by matted fibrous adventitious roots. The core of a mature trunk (Figure 1.19) is composed of a complexly folded vascular cylinder from which originate the numerous veins, passing through the mantle of adventitious roots, to the leaf fronds. Nearer the trunk's crown, the persistent stubs of old fronds (leaves) are also visible and, in Figure 1.20 the apical bud and younger fronds surrounding the unfolding crozier-like new leaves can also be seen. This tree fern is now widely distributed in temperate regions and the one illustrated here was planted many years ago in the grounds of Colonsay House, Outer Hebrides, Scotland.

Tree ferns reproduce sexually from spores, and Figure 1.21 shows the upper surface of a 2 m long fertile frond photographed on the day after its excision from the parent fern, growing in Glasgow Botanic Gardens, Scotland. The dry laboratory atmosphere dried out the frond and a multitude of haploid spores – together with sporangial debris – was liberated from sporangia on the frond's lower surface and left a yellow spore print on the underlying bench. Figure 1.22 shows a fertile male prothallus (the haploid component of the fern's

FIGURE 1.18

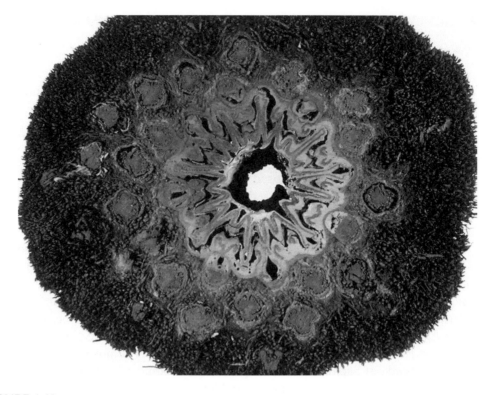

FIGURE 1.19

FIGURE 1.20

FIGURE 1.21

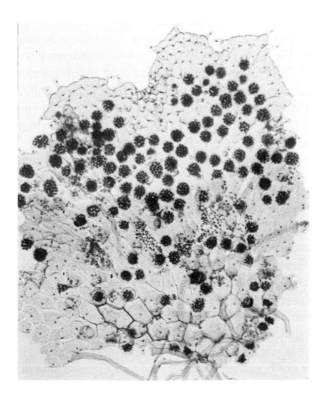

FIGURE 1.22

life cycle) viewed under a light microscope. Such a thin and small (1 cm or so wide) heart-shaped body lacks a protective cuticle and can only survive in humid conditions on a moist substrate. This male specimen exhibits many darkly stained antheridia in which motile flagellate sperm develop; potentially, these would swim in a film of water to fertilise eggs in an adjacent female prothallus. A few of the resulting diploid embryos would grow into mature diploid plants and thus complete the fern's alternation of generation life cycle.

ARBORESCENT SEED PLANTS

Gymnosperms are almost entirely arborescent and comprise over 600 species of conifers, 100 cycads, a single species of *Ginkgo, Ginkgo biloba,* whereas, of the three gnetophyte genera, only *Gnetum gnemon* is a tree. Pollination in the gymnosperms is mostly effected by wind but beetle vectors may be involved in cycads.

Conifers often grow into very large individuals and many are extremely long lived. In the famous Mariposa Grove in the Sierra Nevada of California, United States, there are many giant trees of *Sequoiadendron giganteum* (Wellingtonia), whereas, on the Californian coast, one specimen of *Sequoia sempervirens* (Coastal Redwood) is 111 m tall. Other North American conifers, such as *Abies procera* (Noble fir), may grow to 70 m in height. The oldest, accurately dated living tree is the nearly 4,770-year-old specimen of *Pinus longaeva* (Bristlecone Pine) in the White Mountains of California.

Most conifers are evergreens, with their leaves shed over a few years, whereas each of the large leaves of *Araucaria araucana* (Monkey Puzzle) lasts up to 30 years. Figure 1.23 illustrates a mature cultivated specimen growing in the grounds of Balcary Bay Hotel, Galloway, Scotland. This conifer is indigenous to the western slopes of the Chilean/Argentinian border of the Andes, where it once formed dense forests at altitudes from 1000 m upwards. But overlogging, overgrazing, and extensive forest fires have now caused it to be classified as endangered by the IUCN. It is now widely grown in temperate regions as a landscape feature (Figure 1.23). Probably because of its sheltered environment in Galloway, it is still densely clothed in leafy branches, whereas, in its native Andean habitat, the lower branches of the tree die off, and mature trees usually bear only a terminal cluster of branches. The female tree develops large globular cones (Figure 1.24) and its abundant seeds provide a highly nutritious food for the native Andeans.

The conifer *Pinus sylvestris* (Scots Pine) typically has a reddish-brown bark (Figure 1.25). This species is still widespread in Eurasia, ranging from Western Europe to Eastern Siberia. However, as the climate warmed at the end of the last Ice Age, it became extinct in Britain except in Scotland, where it formed a major element of the Caledonian Forest in the Scottish Highlands. Subsequently, logging, fire, and overgrazing there have led to a drastic reduction in the extent of this forest. Nevertheless, Scots Pine is still prominent in reservations in several Highland glens.

Another conifer, *Juniper communis* (Common Juniper), is shown in Figure 1.26. It is growing on an exposed valley slope in the Lake District, England. This conifer has a wide geographical range, with a circumpolar distribution in the cool temperate Northern Hemisphere. It is very variable in form, and ranges from trees up to 16 m tall to low shrubs, as illustrated here. It is one of the only three conifers which are indigenous to Britain. Cones of the juniper are borne on separate male and female individuals, and the female cones develop into small, one-seeded "fruits". These are sometimes gathered and used to flavour gin or for other culinary purposes.

FIGURE 1.23

FIGURE 1.24

FIGURE 1.25

FIGURE 1.26

Yews are also conifers and this ancient specimen of *Taxus baccata* (Yew tree, Figure 1.27) is situated on the southern bank of the River Gryfe in Renfrewshire, Scotland. At ground level, the circumference of the massive coppice-like Yew stool is some 6.5 m, but, over the centuries, it has given rise, by layering of its lower branches, where they touch the ground, to a magnificent Yew grove. This occupies an area of approximately 40 × 20 m (as measured from a Google aerial map). The origin of the coppice stool remains uncertain. It may have resulted from the death of the main trunk and subsequent branching from its base, or it might have resulted from an original bundle planting of several Yew saplings. DNA sampling from different main boughs of this specimen would probably determine which alternative is correct. Figure 1.28 shows the margin of this Yew grove with the rooting of

FIGURE 1.27

FIGURE 1.28

its layered branches. The closely adjacent River Gryfe is also visible, together with debris deposited in the Yew's branches, resulting from recent flooding.

The massive, squat conifer *Agathis australis* (Tane Mahuta), shown in Figure 1.29 is growing in Waipoua Forest, North Island, New Zealand. This truly majestic tree (sacred to the Maori) is stated to be 51.5 m tall with a girth of 13.8 m; it is probably the oldest-living individual Kauri and it has been estimated to be 2,000 years old. Its crown is nearly 35 m wide, with numerous epiphytes growing on it, and its largest branches exceed 1 m in width. A recent report expressed alarm about the spread of the soil-borne tree pathogen *Phytophthora agathidicida*, which has now spread to within some 60 m of Tan Mahuta. The younger specimen of *Agathis* (Figure 1.30) was photographed on the roadside leading to the Waipoua Forest. Its large monopodial trunk is terminated by a well-developed canopy.

FIGURE 1.29

FIGURE 1.30

FLOWERING PLANTS

The mature specimen of *Quercus petraea* (Sessile Oak) shown in Figure 1.31 is thriving in a small plantation of trees all over one hundred-years-old in an Oak pasture in Kirkconnel, near Dumfries, Scotland. This species is native throughout Europe, but about 600 other species of Oak, including *Quercus robur* (English Oak), have been identified elsewhere in the world. This ancient hollow *Q. petraea* specimen (the Clachan Oak, Figure 1.32) is an ancient pollarded tree, growing in Balfron, Stirlingshire, Scotland. Local legend claims that the famous 18th century outlaw Rob Roy MacGregor hid in the tree. It was recorded in the mid-19th century as then being some 300 years old, and that it had been struck by lightning earlier in that century; this possibly accounts for its present pollard-like form. It seems that the three iron hoops enclosing the hollow trunk were originally installed so that petty miscreants could be left chained to the tree for public ridicule. However, the

FIGURE 1.31

FIGURE 1.32

hoops and wooden staves now in place on the trunk are more recent and intended to stabilise this remarkable tree.

A mature specimen of the tree *Brachychiton rupestris* (Bottle tree) is pictured in Figure 1.33. This drought-resistant member of the Malvaceae has a grossly distended trunk and is growing in the Royal Botanical Gardens, Sydney, Australia. The species is deciduous and flowers in September when it sheds its leaves. The soft pulpy tree wood has been used as emergency cattle feed during droughts in the Australian bush.

A row of *Fagus sylvatica* (Beech tree) borders the roadside in Galloway, Scotland (Figure 1.34). This coastal area is often subject to strong cross winds and gales which result in the windswept appearance of the upper branches of these trees. Figure 1.35 shows a solitary specimen of the angiosperm *Betula pubescens* (Downy Birch tree), which has taken root on Holme Fell in the Lake District, England. It is deciduous and sometimes reaches 20 m in height with a trunk up to 70 cm wide. Its wind-pollinated catkins are produced in early spring, before the leaves. The winged seeds are widely distributed by the wind, as demonstrated here by its lonely position on the hillside moorland.

A repeatedly pollarded specimen of *Salix* (Willow), growing by the river Cam in Cambridge, England, is illustrated in Figure 1.36. This tree is often planted along water courses since its branches root freely and its tangled root system stabilises the soil against mechanical erosion by the flowing water. Willow branches root readily and often cause problems around housing where the roots may clog drainage systems. Flexible young Willow canes (harvested from pollarded stumps) can be employed for basketry, whereas older branches are employed for fencing or other agricultural uses. Most willows readily hybridise, which often makes their specific identification difficult.

This mature specimen (Figure 1.37) of *Ficus macrophylla* (Moreton Bay Fig) was planted in about 1870 by the first Collector of Customs at Russell, North Island, New Zealand. The

FIGURE 1.33

FIGURE 1.34

FIGURE 1.35

FIGURE 1.36

evergreen rainforest tree is native to eastern Australia but is now widely found (as shown here by its location in New Zealand) in other subtropical regions. It may reach 60 m tall with a trunk up to 3 m in diameter. In its rainforest habitat, it usually occurs as an epiphytic "strangler" tree, due to the germination of its animal-distributed seed in a host tree canopy. The growing fig roots fuse as they grow down towards the forest soil and eventually encase the host tree's trunk. The latter eventually dies and leaves the fig as a free-standing tree.

A young specimen of the monocot *Dracaena draco* (Dragon tree), which is now widely planted as a subtropical ornamental tree, is illustrated in Figure 1.38. Its image has now been adopted as the symbol of Tenerife in the Canary Islands. This woody monocot tree, after some 10–15 years of monopodial growth, produces a terminal spike of white flowers. Vegetative growth of the tree then resumes from several axillary buds, as is illustrated here. These branches grow for a further several years until re-branching occurs. This growth habit is progressively repeated to eventually yield an umbrella-like woody tree (Figure 1.39). Due to its unusual form of secondary growth, it does not form annual rings as the trunk thickens. Its age is thus difficult to determine but this specimen is thought to be at least 250 years old.

These monocot specimens (Figure 1.40) of *Xanthorrhoea australis* (Bukkup, Grass tree) are growing in the bush in Queensland, Australia and reach several metres in height. Figure 1.41 illustrates another specimen of *Xanthorrhoea*, showing its tall, spear-like flowering spike with numerous small flowers displayed on it.

A massive fan-like leaf of the monocot tree *Ravenala madagascariensis* (Traveller's Palm) is shown in Figure 1.42. This monocot tree is not a palm but belongs to the

FIGURE 1.37

FIGURE 1.38

FIGURE 1.39

FIGURE 1.40

FIGURE 1.41

FIGURE 1.42

Strelitziaceae family. Its long petioles are terminated by up to 30 paddle-like blades; these all lie in the same plane, which tends to be orientated in an east-west direction – hence its popular name. In some varieties, the tree trunk may grow to about 60 cm in diameter and up to 30 m tall.

SEXUAL REPRODUCTION IN ARBORESCENT SEED PLANTS

Gymnosperms

Figure 1.43 shows the appearance, in late spring, of cylindrical clusters of male cones of *P. sylvestris* (Scots Pine). These cones soon expand and dehisce to liberate vast quantities of wind-blown pollen. Unlike in angiosperms, the subsequent conifer seed is not enclosed by a fruit wall but instead is "naked".

FIGURE 1.43

The earliest fossils of the genus *Cycas* occur in the Cenozoic era. Various trees of *Cycas media* (Figure 1.44) grow in Fletcher Botanic Gardens, Cairns, Australia. This splendid specimen is unfolding a new crop of foliage leaves and bears a massive cluster of large green fruit-like seeds. This genus differs from other cycads in not forming seed cones but instead bears leaf-like megasporophylls on female plants, whereas distinct cones develop on the male trees.

The endangered arborescent cycad *Lepidozamia peroffskyana* (growing in its native habitat in the Tambourine Mountains, SE Queensland, Australia) is shown in Figure 1.45. This large, ancient specimen has a massive terminal cone, probably weighing some 40 kg. The tree is now growing in the garden of a house in the township of Eagle Heights, but this specimen greatly predates the establishment of this town. The tree's fully expanded leaves may reach over 3 m long. Figure 1.46 illustrates a massive female cone of an unidentified cycad growing in a park in Madeira.

Seeds of a mature specimen of the gymnosperm *G. biloba* (Maidenhair tree) are shown in Figure 1.47. The tree is laden with ripe seeds (not fruits) and is growing in Poznan Botanic Gardens, Poland. This gymnosperm still exists in the wild in two small areas of eastern China. However, it has long been cultivated in temple grounds in Korea and Japan, with some trees being 1,500 years old. The tree is deciduous, with distinctive fan-shaped leaves, and trees may reach 50 m tall. *Ginkgo* is a "living fossil" with some leaf remains dating back 270 million years. Female plants do not form cones but instead two ovules develop on a stalk; after pollination, the eggs are fertilised by massive flagellate motile sperm. The embryos develop into small, fleshy fruit-like seeds on female trees but their

FIGURE 1.44

FIGURE 1.45

unpleasant smell means that male plants are often favoured when *Ginkgo* is planted as residential street trees in California and elsewhere.

Figure 1.48 shows a male specimen of *Welwitschia mirabilis* (Kharos), the cones of which are about to shed their pollen. The plants were grown from seed in the Berlin Botanic Gardens, Germany and survived the fighting and aerial bombing of WW2. This monotypic gymnosperm species is endemic to the Namib desert of Namibia and Angola (where it may attain 1.5 m in height), and belongs to one of three extant genera in the Gnetales order of seed plants, whose evolutionary relationship to other seed plants is still uncertain.

FIGURE 1.46

FIGURE 1.47

FIGURE 1.48

FLOWERING PLANT (ANGIOSPERM) TREES

A close-up view of the flower of the basal dicot tree *Magnolia* is illustrated in Figure 1.49. Several of its perianth members were removed to reveal the elongated receptacle which bears numerous carpels in a spiral at its apex and many stamens below. The already-withering anthers occupy the terminal portion of each stamen.

A cluster of young green fruits of the monocot *Cocos nucifera* (Coconut tree) is shown in Figure 1.50. These are developing near the top of the palm tree trunk. The very large, highly nutritious fruit develops from a trilocular ovary, as is evident from the triangular tip of the young fruit. Numerous glistening ripe fruits of the monocot *D. draco* (Dragon tree) are shown in Figure 1.51. These are borne on a tree growing in the Royal Botanic Gardens, Sydney, Australia.

A longitudinal section (as seen under a light microscope) of an immature dicot embryo is shown in Figure 1.52. Its paired cotyledons are already well-developed, and a mound-shaped rudimentary plumule (future shoot) lies between their bases, whilst a large radicle (embryonic root) terminates the other end of the seed. Figure 1.53 shows recently dehisced fruits of the dicot *Salix* (Willow tree). Its abundant hairy seeds are wind blown and become widely distributed in the adjacent landscape.

FIGURE 1.49

FIGURE 1.50

FIGURE 1.51

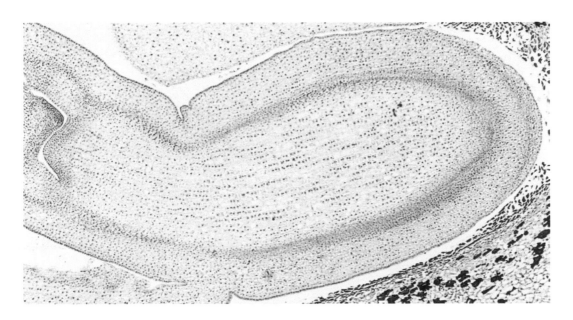

FIGURE 1.52

FIGURE 1.53

2 Anatomy of Trees and Woody Plants

The multicellular growing points of actively growing shoots and roots of several vascular plants are illustrated in Figures 2.1–2.5. In the radial longitudinal section of the vegetative shoot tip of the conifer *Pinus* (Figure 2.1), its hemispherical apex is closely invested by several protective, needle-shaped leaf primordia, which are connected to procambial strands from the young stem below. In Figure 2.2, the vegetative shoot apex of a dicot flowering plant, with decussately arranged leaves, shows the swollen bulges of incipient axillary buds already visible in the axils of its two leaf primordia.

In contrast to the organisation of the shoot, in the root, the meristematic initials in its tip (Figure 2.3) are protected from abrasion from soil particles by its prominent, closely fitting root cap. In this specimen, note the prominent files of enlarged cells, within the central procambial core, which demarcate the potential metaxylem. In the cross-section of a young dicot root (Figure 2.4), several prominent, thick-walled metaxylem elements have differentiated in a triarch pattern from the central procambial cylinder, whereas, on the alternating, radial procambial axes, three islands of phloem are also visible. In the cross-section (Figure 2.5) of a young *Salix* (Willow) root, two lateral root primordia are visible within the main root. These have developed from the parent root's pericyclic tissue, are penetrating its cortex and will emerge as lateral roots into the surrounding soil. Note also the vascular linkage evident between the upper primordium and the parent root.

A cross-section of the primary root of *Pinus* (Figure 2.6) root shows its diarch, red-stained primary xylem core. Its two arms are terminated by large, green-stained resin ducts. Smaller resin canals are also visible within the prominent, radially arranged, and thick-walled secondary xylem elements. The root is bounded externally by remains of the epidermis, whereas well-developed but smaller-celled secondary phloem lies adjacent to the xylem core.

In the conifer *Pinus* (Pine; Figure 2.7), a cross-section of the terminal bud shows that it is bounded by the epidermis. The central pith is surrounded by a distinct ring of discrete primary vascular bundles enclosed within a wide cortex, containing several prominent resin canals.

In contrast to the ring of vascular bundles present in young stems of both conifers (Figure 2.7) and dicots, a cross-section of a young monocot stem (such as *Zea*, Maize; Figure 2.8) reveals a scattered vascular system lying within ground tissue. Within each of these bundles, a prominent strand of phloem is composed of smaller, densely staining companion cells and their associated (apparently empty) larger sieve elements. The endarch xylem strand within each bundle shows several wide-lumened, thick-walled vessels. In a twig of the monocot tree *Dracaena* (Dragon tree; Figure 2.9), its cross-section shows an initially un-thickened organisation in the lower area of the section. However, discrete secondary bundles are formed from a discontinuous vascular cambium in the upper part of the section. In Figure 2.10, the cross-sawn face of the trunk of the monocot *Hyophorbe lagenicaulis* (Bottle Palm tree) shows that the wood of its swollen trunk is composed of many discrete, fibrous vascular strands.

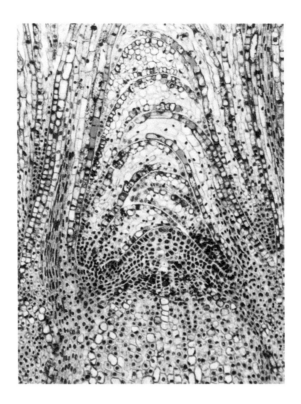

FIGURE 2.1

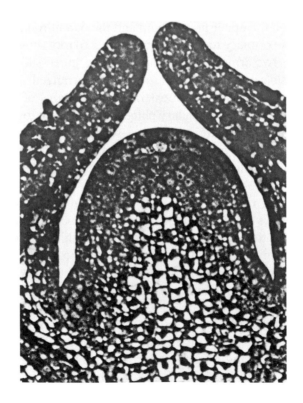

FIGURE 2.2

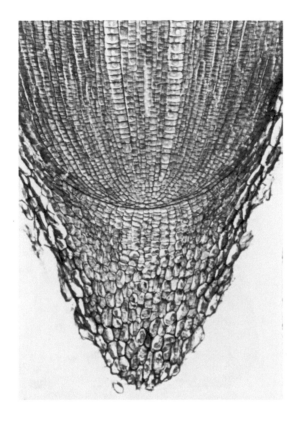

FIGURE 2.3

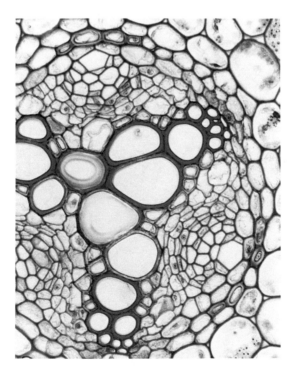

FIGURE 2.4

FIGURE 2.5

FIGURE 2.6

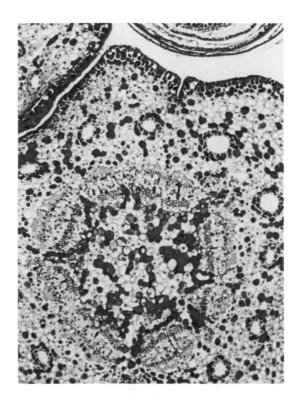

FIGURE 2.7

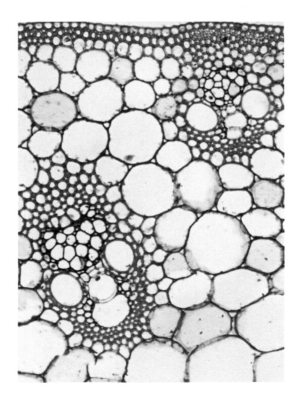

FIGURE 2.8

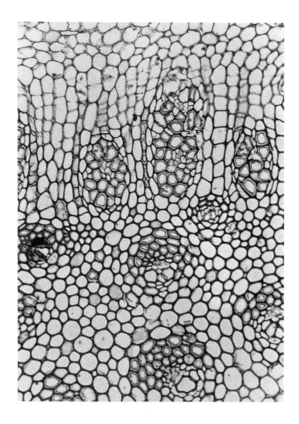

FIGURE 2.9

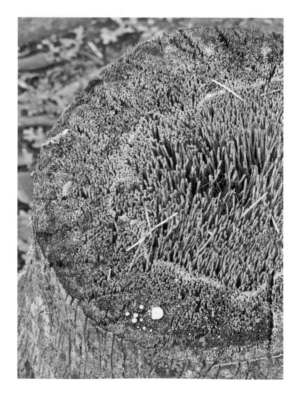

FIGURE 2.10

The primary xylem in vascular plants contains numerous lignified elements and Figure 2.11 shows a longitudinal section of a xylem strand from a young dicot stem. On the left, the protoxylem tracheary elements show annular to spiral and scalariform secondary-wall thickenings. However, on the right, thicker-walled pitted or scalariform metaxylem elements are visible.

A cross-section of a young stem of *Sambucus nigra* (Elderberry) is shown in Figure 2.12. This small deciduous European dicot tree develops prominent lenticels in its bounding layer of cork, as illustrated here at the top of the section. Note also the prominent cylinder of green-stained secondary xylem.

Figure 2.13 shows a fragment of cork from the debarked stem of *Quercus suber* (Cork Oak), showing several of the numerous lenticels which traverse the otherwise impervious cork from the outside atmosphere (black, wavy layer) to the inner, still-living stem tissues. Lenticels facilitate gaseous interchange between the stem (or root) and the external environment.

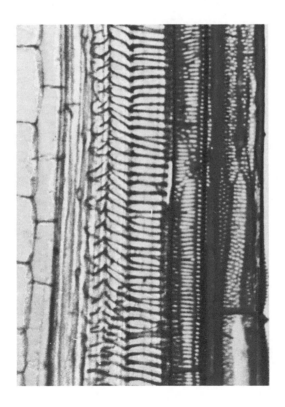

FIGURE 2.11

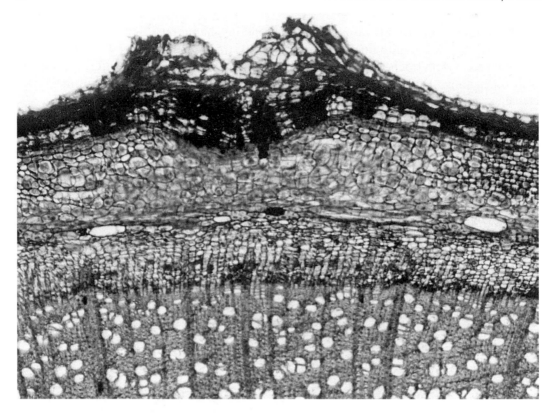

FIGURE 2.12

FIGURE 2.13

SECONDARY GROWTH OF THE STEM

A cross-cut support post to a walkway bridge is shown in Figure 2.14, and its prominent, weathered annual rings show the tree to have been harvested when some 40 years old. In Figure 2.15, the cross-cut trunk of *Eucalyptus* sp. (from the bush in Queensland, Australia) shows its massively thick, fire-resistant, and fibrous outer cork covering.

Figure 2.16 illustrates a cross-cut section of the wood (secondary xylem) of the conifer *Thuja plicata* (Western Red Cedar). It is a long-lived tree, native to the Pacific North West of America, where some specimens exceed 1,300 years of age and may reach 70 m in height, with trunks up to 4 m wide. As with all conifers, vessels are absent from the xylem. Its highly prized timber is resistant to decay and is frequently used outdoors for roof shingles, posts, and decking. Its early wood is composed of radially aligned tracts of thin-walled, polygonal tracheids; however, in the late wood, these become abruptly tangentially flattened and much thicker walled, as is evident in the lower area of the image.

In contrast to conifers, the wood of dicot trees (Figure 2.17) develops both vessels and tracheids, as illustrated in the ring-porous wood (secondary xylem) of *Robinia pseudo-acacia* (False Acacia tree). In a cross-cut section of a twig, numerous parenchymatous, thick-walled rays traverse the wood radially. In the late wood, the vessels are smaller and more closely packed than in the season's early wood (shown in the top and middle part of the section). The early wood contains large, wide-lumened vessels which are packed with

FIGURE 2.14

FIGURE 2.15

tyloses; these are ingrowths from the xylem parenchyma (*via* pits in the walls of adjacent vessels) into the vessel lumens.

Figure 2.18 shows several brightly coloured monocot trunks of *Crytostachys renda* (Sealing Wax Palm) growing in a decorative garden border in Thailand. This slow-growing palm is multi-stemmed, with narrow stems up to 10 cm wide. It can reach some 10 m tall in the peat-swamp forests of SE Asia. Its common name is due to the bright red colouration of its crown and leaf base sheaths, resembling the colour of sealing wax.

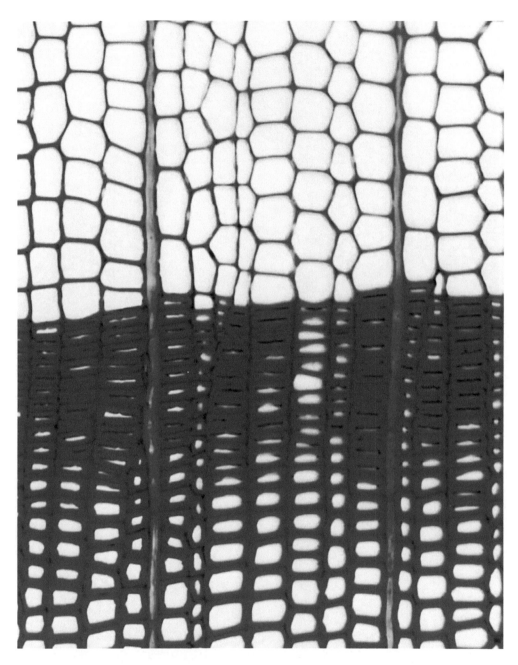

FIGURE 2.16

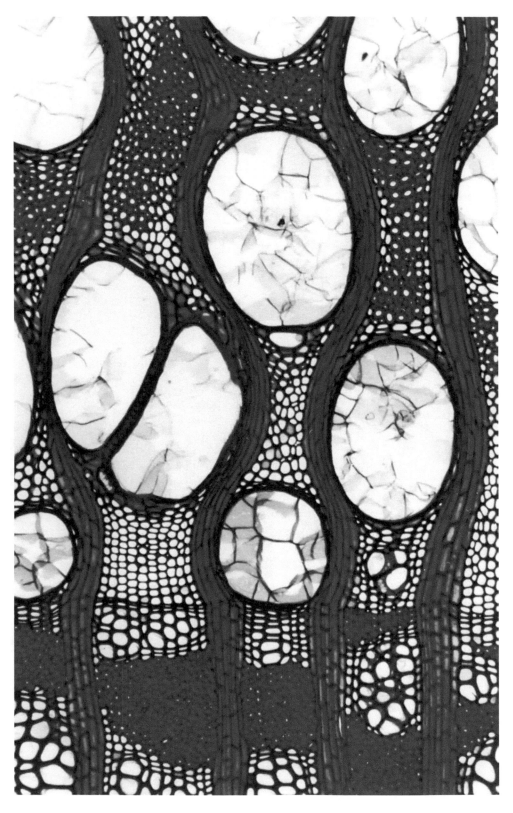

FIGURE 2.17

FIGURE 2.18

MODIFICATIONS OF TREE BARK

In Australia, there are at least 900 species of *Eucalyptus*. Figure 2.19 illustrates the lower trunk of *Eucalyptus nitens* (Shining Gum tree), which is native to wet forest areas of eastern Australia. As shown here, its semi-persistent original grey, flaky bark is later shed from the trunk in ribbons. The trunk reaches 60 m in height and is an important plantation timber tree in Tasmania.

FIGURE 2.19

FIGURE 2.20

FIGURE 2.21

There are some 170 species of the dicot *Banksia* present in Australia, and this specimen of *Banksia serrata* (Saw-Toothed Banksia; Figure 2.20) was growing in Ku-ring-gai Wildflower Garden, near Sydney. It usually has a single knobbly, grey trunk with a warty bark (cf. Figure 2.21) several centimetres thick. The tree grows up to 15 m high, with shiny, dark green, serrated leaves (hence its common name) and, in summer, develops large, yellow flower spikes. The tree is often engulfed in bush fires; however, it survives the flames by sprouting from epicormic buds protected under its thick bark. Specimens of this tree were first collected for scientific study by Joseph Banks and Daniel Solander on 29 April 1770 in Botany Bay, Australia; these naturalists accompanied James Cook on his first voyage to the Pacific. The cross-cut surface of a small branch of *B. serrata* is shown in Figure 2.21. Note its deeply incised, warty bark overlying a narrow cylinder of secondary phloem, which encloses a prominent core of secondary xylem. The older bark eventually exfoliates in reddish-orange flakes to reveal a smooth, whitish new bark surface to the trunk.

The trunk of *Betula dahuvica* (Asian Black Birch) is shown in Figure 2.22. The tree is deciduous and may grow to some 13 m in height. Its grey-brown bark becomes fissured and exfoliates in paper-like curls.

The large deciduous tree *Aesculus hippocastanum* (Horse Chestnut) was introduced to Britain from Turkey in the second part of the 16th century and is now widely planted in the UK and in Europe as an amenity tree. Deer and other mammals eat the abundant supply of food provided by the fall of its fruits ("conkers") in autumn. The trunk of the young tree appears smooth and pinkish-grey but the old bark often reveals a deeply incised plate-like exfoliating surface (Figure 2.23).

FIGURE 2.22

FIGURE 2.23

FIGURE 2.24

Arbutus andrachne (Greek Strawberry tree; Figure 2.24) is a small, evergreen erica-ceous tree native to Southern Europe and the Middle East, and sometimes attains 10 m in height. During summer, its bark flakes into curly red strips which exfoliate to reveal under-neath a smooth creamy trunk.

Part of a young branch from a large specimen of the deciduous tree *Ceiba speciosa* (Silk Floss tree) is shown in Figure 2.25. It is densely covered with thick, very sharp, and conical thorns which help to deter foraging animals.

Sharp thorns are also formed on the trunk of *Gleditsia sinensis* (Chinese Honey Locust tree; Figure 2.26). This deciduous leguminous Asian tree may grow to 16 m in height. Its thorns have been used for over 2,000 years in traditional Chines medicine, and recently thorn extracts have been reported to exhibit anti-tumour properties for the treatment of colon cancer.

Eucalyptus haemostoma (Scribbly Gum tree) is a small, smooth-barked tree, with this specimen growing in the bush near Sydney, Australia (Figure 2.27). The tracks in its bark represent tunnels made by the larvae of *Ogmograptis scribula* (Scribbly Gum Moth), which had earlier laid its eggs in the outer layers of the older bark. Later, the newly hatched larvae chew into the bark and the resultant tunnel complexes are shown up when the older bark falls off. The widths of the tunnels gradually increase as the larvae feed on the bark and the tracks terminate where the larvae have pupated.

Figure 2.28 shows the base of a clump of the tropical woody monocot grass *Bambusa tulda* (Indian Timber Bamboo). Here, its green, closely crowded culms have been decorated

FIGURE 2.25

by young lovers. At maturity, each culm can expand to some 8 cm in diameter and reach some 20 m in height. This Asian bamboo is now widely grown in South America and elsewhere. It can be used in the production of paper pulp, whilst its strong, solid culms provide excellent timber for scaffolding and general construction.

The frame of an old bike (originally fastened to the trunk of a specimen of *Acer pseudoplatanus*, Sycamore) has become embedded in the tree by its expanding bark (Figure 2.29).

FIGURE 2.26

FIGURE 2.27

FIGURE 2.28

This specimen is located at Brig o' Turk, Stirlingshire, Scotland, and the tree is estimated to be some 120–150 years old. One limb of its coppiced trunk bears various pieces of embedded metal, which gave rise to its popular name as "The Bicycle Tree".

Figure 2.30 shows the trunk of *Aesculus hippocastanum* (Horse Chestnut), growing in Mugdock Country Park, near Glasgow, Scotland. It shows an occluded, corky lateral branch scar, resulting from the earlier removal of a small twig; on its scar, a series of concentric rings are evident (a total of 18 can be counted), which probably indicates the number of years since the originally pruning.

On the mature trunk of *Fagus sylvatica* (European Beech), an old side-branch scar is visible (Figure 2.31). A prominent wound collar surrounds the now badly decayed xylem core of the excised branch. Figure 2.32 shows the debarked surface of a large Fagus Sylvatica (Beech tree) trunk with the scars of a witches broom visible.

The stump of a mature roadside *Populus* sp. (Poplar tree) is shown in Figure 2.33. This was felled at ground level and numerous epicormic buds are developing from its proliferating vascular cambial tissue.

FIGURE 2.29

FIGURE 2.30

FIGURE 2.31

FIGURE 2.32

A cross-cut trunk of a felled specimen of the conifer *Araucaria araucana* (Monkey Puzzle) is illustrated in Figure 2.34. This mature specimen was part of a small plantation in Castle Kennedy Gardens, Galloway, Scotland. Streams of resin (now congealed) have exuded from the severed resin canals, located in the secondary phloem of its trunk.

FIGURE 2.33

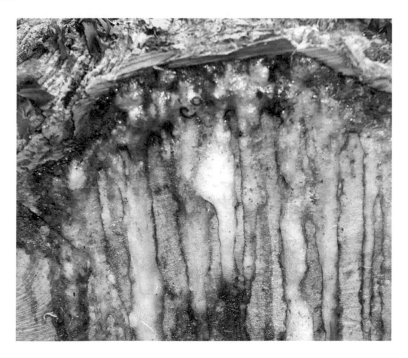

FIGURE 2.34

3 Roots of Trees and Woody Plants

The basic function of tree roots is to facilitate the uptake of water, with its dissolved mineral nutrients, whilst also providing anchorage, and sometimes support, in the soil where the tree is growing. Roots also supply plant hormones to the shoot system, often store starch and may be modified in various ways, such as with breathing roots, for perennation and for nitrogen fixation.

In most trees, their original vertical taproots, which developed from their embryonic seed radicles, die off to be replaced by a network of superficial feeder roots. This veteran pollarded specimen of *Quercus petraea* (Sessile Oak; Figure 3.1) is growing on the western shoreline pasture of Loch Tay, near Killin, Scotland. The tree is growing on level ground, which is often extensively flooded. The surrounding ground has been greatly denuded of its soil to reveal the ring of mechanically stabilising, thickened roots which branch (this event occurring out of sight, underground) and spread outwards as feeder roots. According to most foresters, it is likely that the lateral reach of the feeder roots from such a large tree would be at least equal to the height of its trunk.

Figure 3.2 shows a side branch of a young specimen of *Salix* (Willow) growing in the shoreline mud of a site in Queensland, Australia, where it has readily rooted. Willow cuttings were introduced from Britain in the early days of the then-British colony but the tree is now classified as a non-indigenous plant pest in Australia. As is evident from this image, the willow shoots root easily and a new individual tree will soon develop.

On this specimen of *Ficus elastica* (Rubber tree, Figure 3.3), growing in Malaga (Spain), a large aerial root has grown down from a tree branch to the soil below where it has proliferated to form a strong and supportive buttress root system. Another specimen of *F. elastica* (Figure 3.4; in Tenerife, Canary Islands) has developed abundant adventitious roots from a tree bough, some of which may eventually form supportive pillar roots. Figure 3.5 shows a massive (unidentified) riverside tree growing near Tortuguero at the edge of the jungle in the tropical climate of the Caribbean coast of Costa Rica. From its huge trunk, it has formed large buttress roots which extend into the muddy soil of the adjacent river. Elsewhere in Costa Rica (near Monteverde), the stump of a logged, unidentified rainforest tree (Figure 3.6) reveals its hollow trunk, bearing several buttress roots, with that on the left having split lengthwise into two components.

The exposed root system of *Pinus nigra* (Corsican Pine, growing on an eroded coastal sand dune in Norfolk, England) is shown in Figure 3.7. Note the very considerable lateral extension of several of its main roots, which have helped to stabilise the tree trunk in its precarious sandy location. Figure 3.8 illustrates the root system of *Pinus sylvestris* (Scots Pine) following erosion, growing on the sandy shore of Loch Lomond, Scotland. Scots Pine may develop a deep taproot or a shallow root system, depending on the habitat, so that, consequently, it is often used for the reforestation of former industrial sites.

Trees often grow in apparently inhospitable sites (cf. Figure 3.7), as is well illustrated by this specimen of *Fagus sylvatica* (Beech tree; Figure 3.9), growing on the sheer face

FIGURE 3.1

FIGURE 3.2

FIGURE 3.3

FIGURE 3.4

FIGURE 3.5

FIGURE 3.6

FIGURE 3.7

FIGURE 3.8

FIGURE 3.9

of an abandoned stone quarry located by the Mugdock Reservoir in Milngavie, near Glasgow, Scotland. Note how the tree has spread from its initial site about one-third of the way up the quarry face, where a chance beech seed had germinated, but its roots have now traversed several rocky crevices to anchor the small tree trunk to its precarious vertical rock site.

Figure 3.10 shows a specimen of *Rhizophora* sp. growing on the intertidal shoreline at Cape Tribulation, Queensland, Australia. As with other trees subjected to regular tidal inundation by the sea, its root system shows various adaptions to the salty environment, as is demonstrated in this close-up image (Figure 3.11) of its tangled prop root system. In other mangrove species, such as *Avicennia marina* (Grey Mangrove), growing in waterlogged swamps (Figure 3.12), the roots are modified to form pneumatophores ("breathing roots"); these are richly covered with lenticels through which gaseous exchange occurs when the roots are not submerged by the tide.

Palm trees are monocots which develop an anchoring fibrous root system from their trunks near ground level; Figure 3.13 shows numerous young adventitious root tips penetrating the bark of an (unidentified) Australian palm. The lower trunk of an abundantly rooted specimen of the Australian monocot shrub *Pandanus* sp. is illustrated in Figure 3.14. In this cross-section of the lower trunk of an unidentified Costa Rican palm tree (Figure 3.15), numerous scattered vascular bundles are evident, with a tangled mass of attached woody adventitious roots being visible in the upper area of the section.

FIGURE 3.10

FIGURE 3.11

FIGURE 3.12

FIGURE 3.13

FIGURE 3.14

FIGURE 3.15

Trees are frequently severely buffeted by gales. In this wind-felled specimen of *F. sylvatica*, Figure 3.16), the lateral feeder root system was not sufficiently secured by the soil on the steeply sloping bank of the mound where it had been planted.

Various alternative types of root systems are displayed by other tree species. In the palm *Socratea exorrhiza* (Walking Palm or Cashapona), found in the forest near Arenal, Costa Rica (Figure 3.17), it seems that, if the palm collapses, due to predation by grazing feral pigs or other forest animals (note several bitten-off root tips shown in this figure), then the stem can right itself vertically (but slightly to one side of its original position), due to the elongation of newly-grown adventitious roots. Hence, this palm can be said to "walk".

In Figure 3.18, the trunk of a forest tree in Queensland, Australia has been invested by the adventitious roots of *Ficus watkinsiana* (Strangler Fig), which fuse together around the host tree, which eventually dies from the restriction of its water supply and lack of sunlight. Sometimes, the hollow central space (originally occupied by the now-dead and rotted trunk of the parasite's host tree) can be viewed, as is evident in this internal view (Figure 3.19) of a hollow strangler fig, showing its fused roots.

It is quite common to find adventitious roots developing from wounded forest trees, as is well illustrated by this specimen of an ancient pollarded *F. sylvatica* beech tree (Figure 3.20). Several swollen adventitious roots have formed around the rim of the hollow core to the pollarded trunk, and it is likely that these will eventually grow down the hollow trunk, feeding from the nutrients released from its rotting xylem.

FIGURE 3.16

FIGURE 3.17

As well as lengthening, a growing root expands in diameter and consequently exerts considerable lateral pressure against the soil substratum or any obstructions in its path. Such growth can be very damaging to natural features (helping to split rocks apart) and to man-made objects, such as roads, pavements, and buildings. In the remains of Craigend Castle (at Mugdock Country Park, near Glasgow, Scotland), a chance sapling of *Sorbus aucuparia* (Rowan tree; Figure 3.21) has taken root on its crumbling tower and the expanding roots have gradually forced apart its massive stone blocks. As with the shoot, such aerial roots are well supplied with lenticels (Figure 3.22).

In the wild, the root systems of trees growing closely together often co-mingle. Figure 3.23 shows the intermingled root systems of *Quercus* (Oak) and *Acer pseudoplatanus* (Sycamore) growing on a lakeside shoreline in Cumbria, England. Although it is difficult to trace their individual identities, actual root grafting between separate, but adjacent, individuals of different tree species does not seem to occur in nature.

When preparing areas of wild landscape for estate planting, the persistent remains of root balls from grubbed-up trees and shrubs (Figure 3.24) often need disposal. An excellent example of turning such necessity into an unusual and attractive landscape is provided by the Oak stumpery (Figure 3.25), made from the upended stumps of the many Oak trees which had been cleared from the developing Biddulph Gardens (now a famous National Trust site in Staffordshire, England) as it was constructed at the behest of James Bateman, a visionary 19th century coal magnate.

FIGURE 3.18

FIGURE 3.19

FIGURE 3.20

FIGURE 3.21

FIGURE 3.22

FIGURE 3.23

FIGURE 3.24

FIGURE 3.25

The underground root systems of trees are usually hidden from casual inspection, but sometimes erosion of the overlying soil allows their examination. In Figure 3.26, the soil covering the roots of an *Alnus glutinosa* (Alder tree) specimen from the banks of Loch Lomond, Scotland) had been washed away to reveal this very large nitrogen-fixing nodule growing on its rootstock. Figure 3.27 shows a 28-day-old cultured root nodule, bearing several swollen nitrogen-fixing root tips, of the native American shrub *Comptonia peregrina* (Sweet Fern).

FIGURE 3.26

FIGURE 3.27

4 Tree Leaves
The Green Sugar Factory

The leaves of vascular plants, although their mature size, form, and roles in the adult plant are variable, have a common origin from the margins of the shoot apex (Figures 2.1–2.2). Green leaves play a vital role in photosynthesis, their mesophyll cells being packed with the chloroplasts (Figures 1.1–1.2, Figure 4.1), which are essential for photosynthesis. The leaf stomata (Figures 4.2–4.3) can open and close to regulate the exchange of carbon dioxide, oxygen, and water vapour between the internal and external environments of the leaf.

A range of green angiosperm tree leaves is illustrated in Figures 4.4–4.11. Scale leaves often surround and protect immature foliage leaves within the growth bud. Many dicot trees are deciduous and unfold fresh leaves each year, with a familiar example in Europe being the bursting in spring of the large sticky buds on the tree *Aesculus hippocastanum* (Horse Chestnut; Figure 4.4). This species is native to Turkey but it is now well-established as an amenity tree in western Europe and elsewhere. At the base of the now-swollen bud (Figure 4.4), its sticky scale leaves have been displaced (and later fall off) as a result of the swelling of the new foliage leaves. These soon expand into a green leaf canopy (Figure 4.5), composed of decussately arranged, compound leaves. In autumn, the green foliage leaves senesce, changing their colour from deep green to a mottled patchwork of yellow and reddish-brown shrivelled blades (Figure 4.6), with the senescent leaves falling soon afterwards.

A leafy shoot of *Cecropia* (Yarumo), a woody dicot genus with numerous species in the tropical rainforest of Central America, is shown in Figure 4.7. It is easily recognised in the forest by its large, circular, and umbrella-like leaves, covered by a mat of fine white hairs on their lower surfaces. The tree usually bears only a few branches and may grow to 15 m in height. *Alluaudia procera* (Madagascar Ocatillo; Figure 4.8) is a succulent, water-storing dicot tree, indigenous to Madagascar. It bears stout solitary spines, which may attain several centimetres in length. Its cylindrical bole (up to 0.5 m in diameter) may grow 15 m tall and is often unbranched over its lower reaches. During prolonged dry seasons, it loses its small green leaves. The tree is an important natural source of building wood and fuel and is commonly harvested from the wild by local people. Figure 4.9 shows the shiny foliage of the dicot *Firmiana malayana* (Mata Lembu), which is a deciduous tropical forest tree, growing to some 20 m in height and native to Malaysia. It sheds its leaves after a dry period and sometimes remains bare for several months. During this period, its orange-coloured flowers form on the bare branches and the flower nectar attracts numerous birds, which effect pollination.

The Australian monocot tree *Xanthorrhoea preissia* (Grass tree, Balga; Figure 4.10) forms numerous long cylindrical green grass-like leaves from the apex of its cylindrical woody stem, which may attain several metres in height. *Roystonea oleracea* (Royal or Cabbage Palm; Figure 4.11) is a very tall palm indigenous to the Caribbean but is here shown growing in Brisbane Botanic Gardens, Australia. Its cylindrical, greyish-white trunk reaches up to 40 m in height and is approximately 0.6 m wide. The trunk is terminated by a crown of twenty or so very large pinnate leaves borne on petioles up to 1 m long. The wood can be used in construction, and its sap is fermented to make an alcoholic drink.

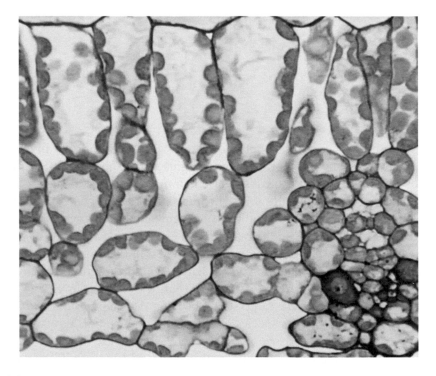

FIGURE 4.1

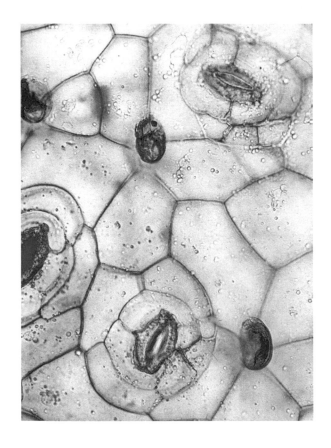

FIGURE 4.2

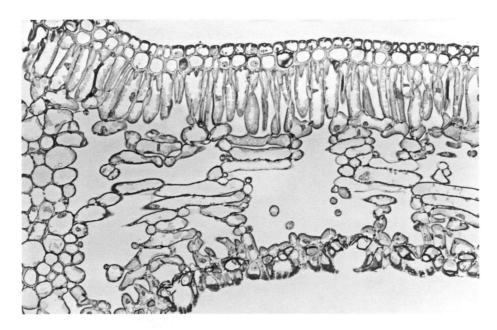

FIGURE 4.3

FIGURE 4.4

FIGURE 4.5

FIGURE 4.6

FIGURE 4.7

FIGURE 4.8

FIGURE 4.9

FIGURE 4.10

FIGURE 4.11

Figure 4.12 shows the outline of several fan-palm leaves growing on a small palm in the Daintree rainforest in Queensland, Australia, whereas Figure 4.13 illustrates a senescent leaf still attached to another fan-palm in the same area of forest. On the King palm, *Archontophoenix* (Figure 4.14), a large senescent pinnate leaf is hanging from the crown of the palm, growing near Darwin, Northern Territory, Australia.

In tropical rainforests, the vegetation is often subject to heavy and prolonged rainfall, and, in some rainforest tree species, their leaves are extended to form drip tips (Figure 4.15). Some biologists have suggested that tropical trees with such drip-tipped leaves, have evolved them to expedite water run-off from leaf surfaces, in order to minimise infection from surface pathogens. However, this theory does not appear to be confirmed by any experimental data. The leaf in Figure 4.15 has been treated with a caustic cleaning agent to remove the cytoplasmic contents, including the pigments, from its mesophyll cells, but to show clearly the network of its anastomosing veins. Figure 4.16 shows in detail the blindly ending terminations of the vein networks in the leaf mesophyll.

Figure 4.17 shows the unfurling of a new fern crozier from the Costa Rican tree fern *Sphaeropteris*; note also the mature leaves from previous growth flushes of this pteridophyte. In the evergreen gymnosperm *Cycas rumphii* (Queen Sago Palm), the new season's crop of large pinnate evergreen foliage leaves is illustrated in Figure 4.18; these are surrounded by older green leaves subtended by small brown scale leaves. Most conifers are evergreen, as exemplified by the spiny-leaved *Araucaria araucana* (Monkey Puzzle; Figure 1.23, Figure 4.19), the individual leaves of which may live for some 30 years. A few conifers are deciduous, such as *Metasequoia glyptostroboides* (Figure 4.20), which loses its foliage in autumn.

FIGURE 4.12

FIGURE 4.13

FIGURE 4.14

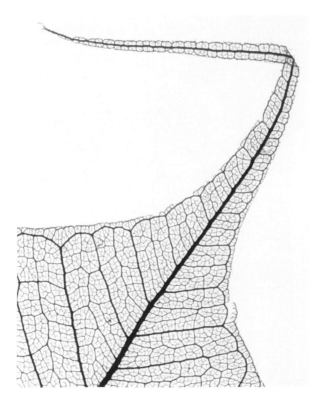

FIGURE 4.15

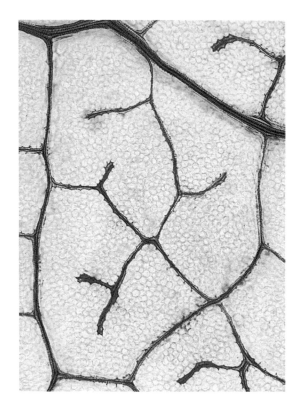

FIGURE 4.16

FIGURE 4.17

FIGURE 4.18

FIGURE 4.19

FIGURE 4.20

5 Reproduction in Trees and Woody Plants

VEGETATIVE REGENERATION

Vegetative regeneration is a common strategy occurring in trees and shrubs, as is illustrated by the proliferation of adventitious shoots from the rootstock of the coastal, fire-tolerant conifer *Sequoia sempervirens* (Coastal Redwood). This tree is native to the Pacific coastal area of SW Oregon and NW California, and when a redwood forest is destroyed by fire, it commonly regenerates from root sprouts (Figure 5.1). As a consequence, the trees grow in clonal clumps, which are supplied with carbohydrates, water, and nutrients from their associated canopy trees, allowing regenerating shoots to grow even in deep canopy shade.

Many woody plants show considerable regenerative capacity, with a protective callus frequently developing over the damaged surface of the plant. This callus often regenerates *de novo* organs, as illustrated by the leafy shoots proliferating on a felled trunk of *Populus* sp. (Poplar; Figure 2.33).

FIGURE 5.1

In *Salix* sp. (Willow), cane cuttings are frequently planted on the margins of lakes, where they quickly take root to form small bushes, which help to stabilise the shoreline (Figure 3.2). The fleshy monocot *Musa* sp. (Banana, Plantain) is not usually considered to be a tree, since its tall aerial stem dies back at the end of each tropical growing season. However, a new trunk arises from a basal bud on the old trunk (Figure 5.2) and grows up to several metres in height before producing a heavy crop of edible fruit (bananas or plantains, Figure 5.3), a phenomenon which is of great importance to tropical economies.

FIGURE 5.2

FIGURE 5.3

SEXUAL REPRODUCTION

The details of ovule development and subsequent maturation in flowering plants and gymnosperms differ considerably (Bowes, "A Color Atlas of Plant Structure", Bryan G. Bowes, 1996, Manson Publishing Ltd., in his Figures 1.32 A–C). In gymnosperms, the ovule is naked and the embryo (as in *Pinus*; Figure 5.4) remains exposed on the tree during its subsequent maturation. By contrast, in flowering plants, the fertilized ovule and developing embryo (Figure 5.5) remain hidden within the carpel. These enlarge and develop into a variety of fruits: Figure 5.6 Dehisced fruits of *Aesculus hippocastanum* (Horse Chestnut); Figure 5.7 Dissected fruit of *A. hippocastanum*; Figure 5.8 Young fruits of *Cocos nucifera* (Coconut Palm); Figure 5.9 Bunch of ripe *C. nucifera* fruits; Figure 5.10 Ripe fruits of *Malus* (Apple); Figure 5.11 Fruits of *Cercis* (Redbud tree); Figure 5.12 Ripe fruits (acorns) of *Quercus* (Oak); Figures 5.13–5.14 Flowers of *Rhododendron* (Rhododendron).

Most angiosperm trees bear bisexual flowers (Figures 5.13–5.14) but some develop unisexual flowers, such as *Salix* (Pussy Willow; Figure 5.15).

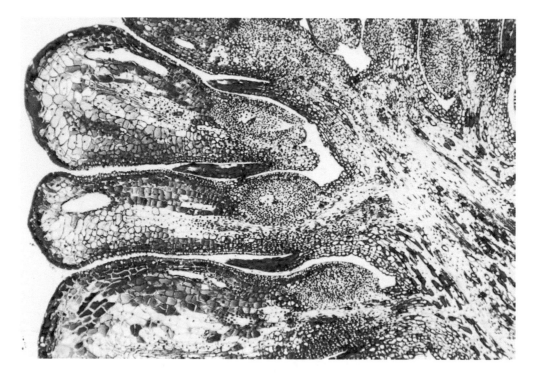

FIGURE 5.4

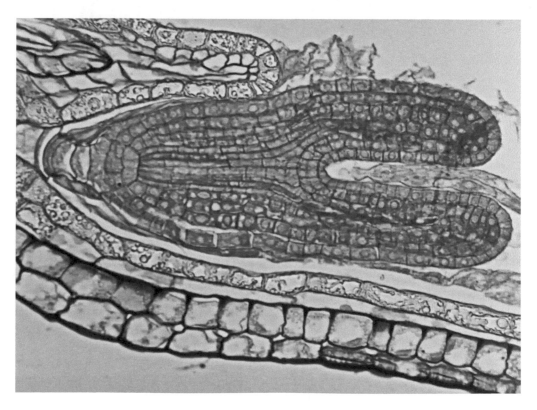

FIGURE 5.5

FIGURE 5.6

FIGURE 5.7

FIGURE 5.8

FIGURE 5.9

FIGURE 5.10

FIGURE 5.11

FIGURE 5.12

FIGURE 5.13

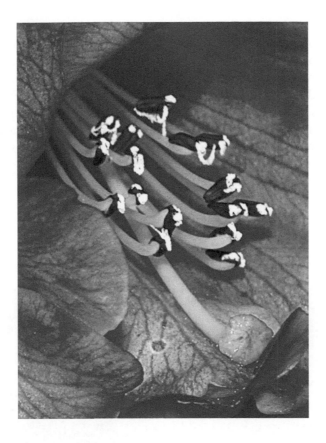

FIGURE 5.14

FIGURE 5.15

The partly-opened blossom of *Liriodendron tulipifera* (Tulip tree, Figure 5.16) provides an excellent example to illustrate the structure of a primitive (basal) angiosperm flower. This spectacular flower develops on a tree which is indigenous to the eastern United States. It grows rapidly, and can reach nearly 50 m in height, with a trunk up to 2 m wide. The tree has been introduced to some British locations (such as Wimpole Farm, National Trust, England), where its large, simple heart-shaped leaves are easily recognisable. In late Spring it produces very conspicuous, large greenish-yellow, cup-shaped, solitary flowers. Its three reflexed sepals soon fall off to reveal a corolla composed of two whorls of alternating large petals. Numerous stamens occur at the base of the floral receptacle and many carpels are inserted on the receptacle above them.

FIGURE 5.16

In some gymnosperms, the sexual organs are borne in massive cones, as shown in the tropical Australian female cycads *Macrozamia fawcettii* (Figure 5.17) and *Lepidozamia peroffskyana* (Figure 5.18), the large, brightly coloured ovules of which are highly toxic for human consumption. In conifers, the female cones are also sometimes large and conspicuous (Figures 5.19–5.21) and their abundant seeds are very attractive food sources for woodland animals (Figure 5.20). The male cones of the gymnosperms are laden with pollen microspores, as illustrated in the longitudinal section of a young *Pinus* cone (Figure 5.22). In Figure 5.23 the swollen, nearly mature, male cones of *Taxus baccata* (Yew) are clearly visible through their translucent distended epidermal covering.

FIGURE 5.17

FIGURE 5.18

FIGURE 5.19

FIGURE 5.20

FIGURE 5.21

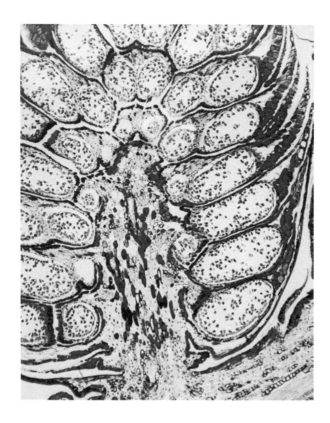

FIGURE 5.22

FIGURE 5.23

6 The Importance of Trees in the Environment

Green trees are vitally important to the environment in which mankind lives, and they are the largest plants to have evolved over the time since land plants first appeared on this planet (cf. Chapter 1). As a result of their photosynthesis, trees liberate oxygen and store carbon in their wood and other tissues, while their root systems stabilise the soil in which they grow. Trees also help provide humans with food and fibres from which to construct habitable shelters in which they may flourish as communities. Trees are also vitally important for wildlife and it has been estimated that a single large and mature European tree may be home to several hundred birds, insects, and small mammals.

In the Blue Mountains of Australia, the heavily weathered sandstone peaks of the Three Sisters (Figure 6.1) tower above the extensive eucalypt forest below. A blue haze blankets the region due to the eucalypts emitting a huge amount of volatile oils from their leaves; when these oils, dust particles, and water vapour combine in the atmosphere, the characteristic blue haze is produced. A closer view of the various trees in this location (Figure 6.2) shows that they are predominantly *Eucalyptus oreades* (Blue Mountain Ash), together with other *Eucalyptus* species. *E. oreades* can grow to some 14 m at maturity; however, unlike most other *Eucalyptus* species, young trees of this species often succumb to bush fires.

Wistman's Wood on Dartmoor in southwest England (Figures 6.3–6.4) is thought to be a relict of the ancient oak forest which covered the region before Mesolithic hunter-gatherers had cleared most of the area by ca 5000 BC. It is one of the highest oak woods situated in the UK and is located on a sheltered, south-west facing slope. Here, a bank of large granite boulders (cf. Figure 6.3) is exposed, revealing pockets of free-draining acidic brown soils. There is no active human management of the site, and cattle and sheep roam freely wherever the steep, boulder-ridden terrain allows. Many people visit the area *via* a track way leading to the southern end of the Wood. The trees growing here are mainly *Quercus robur* (English Oak), but occasional specimens of *Sorbus aucuparia* (Rowan tree), *Ilex aquifolium* (Holly), *Crataegus monogyna* (Hawthorn) and *Corylus avellana* (Hazel) also occur. The tree branches are characteristically festooned with a variety of mosses and lichens (cf. Figure 6.4). On the ground, the numerous large boulders are usually covered by lichens (cf. Figure 6.3) and patches of moss. Wistman's Wood has changed considerably as the climate has generally become warmer. The older oak trees have changed form, from a stunted, semi-prostrate habit to a more upright form, and the new generation of mostly straight-growing, single-stemmed oaks have grown to an average height of approximately 7 m.

The Krummholz conifers shown in Figure 6.5 were growing at Libby Flats, Wyoming. The term "Krummholz" is derived from a German word and describes a stunted, deformed vegetation, where the trees have been shaped by continual exposure to fierce, freezing winds. Under these conditions, trees can survive only where they are partially shielded from the blasting wind by rocky outcrops or covering snow. As the lower portions of such trees continue to develop, their foliage becomes extremely dense near the ground. However, as clearly shown in this illustration, the branches on exposed sides have been killed off by the constant strong freezing winds, thus giving the trees a characteristic flag-like appearance.

FIGURE 6.1

FIGURE 6.2

FIGURE 6.3

FIGURE 6.4

FIGURE 6.5

In Figure 6.6, the wind-sculpted outline of *Acer pseudoplatanus* (Sycamore tree) is shown; this specimen is growing in an exposed hillside garden at Durness, Sutherland, Scotland.

Pneumatophores (breathing roots; Figure 6.7) are abundant on trees of *Avicennia marina* (Grey Mangrove) exposed at low tide at Olympic Park, Sydney, Australia. This species of mangrove has a very wide coastal and tropical distribution from East Africa to SE Asia and Australia. Grey Mangroves grow as shrubs or trees up to approximately 10 m in height, with a smooth light-grey bark. Its large and fleshy seeds often germinate on the tree and fall onto the tidal mud as a juvenile seedling. Its thick glossy leaves are bright green on their upper surfaces. However, the leaves appear silvery-white to grey on their lower surfaces due to deposits of surplus salt excreted from the plant. As with other Mangrove species, its breathing roots grow to a height of up to 1 m, with a diameter of a centimetre or more. The roots are abundantly supplied with lenticels which, when the root is exposed, allow the otherwise submerged root system to absorb oxygen which is lacking in its anaerobic muddy substrate. These roots also anchor the plant, during the frequent inundation with seawater, into the soft substrate of the tidal systems.

The skeleton of this derelict boat, found near the shore in Homebush Bay on the Parramatta River, has been engulfed by *Avicennia marina* (Grey Mangrove; Figure 6.8), which has spread from the adjacent mangrove swamp close to Olympic Park, Sydney, Australia.

In Figure 6.9, the ruined tower of Craigend Castle (in Mugdock Country Park, near Glasgow, Scotland) shows several conifers growing vigorously on the sparse "soil" which has accumulated around the derelict roof surface.

FIGURE 6.6

FIGURE 6.7

FIGURE 6.8

FIGURE 6.9

This derelict railway bridge (Figure 6.10) occurs above a main road in Clydeside, Glasgow, Scotland. The track on this bridge was abandoned some years ago, but the clinker track bed has now become colonised by vigorously growing *Betula* sp. (Birch trees).

The rocky, barren limestone pavement (karst) of the Burren National Park, County Clare, in southwest Ireland, is partially clothed with low-lying, sheep-nibbled vegetation, such as the close-cropped specimens of *Prunus spinosa* (Sloe; Figure 6.11) and *Corylus avellana* (Hazel; Figure 6.12) shown here. In other, more inaccessible, and hence less-grazed, areas of the Burren, these species may grow more fully into bushes.

In Figure 6.13, a specimen of *Xanthorrhoea australis* (Grass tree) has survived a recent bush fire at the Wildflower Garden, near Sydney, Australia. Although this specimen is still quite small, this species grows very slowly and quite small specimens may be several hundred years old. Its trunk, sometimes branched, can grow to several meters tall (cf. Figure 1.41). The wiry, long leaves crown the trunk in a crowded tuft (up to 30 cm long) of needle-like, blue-green leaves. On the older trunks, the leaves become spread out, with the dead leaves forming a skirt around it (cf. Figure 1.41). Aboriginal Australians sometimes eat the soft white crown of this tree (Baggup, Kawee) for food, whilst soaking the flowering spike in water to produce a sweet drink. The flower spike exudes resin which was used as an adhesive, whereas the stem was used as a spear shaft.

A bushfire-blackened specimen of the flower cone of *Banksia serrata* (Saw Banksia; cf. Figure 2.20) is shown in Figure 6.14. Old flower spikes develop into "cones" of up to thirty wrinkled dry fruits, with the withered flower remains giving them a hairy appearance.

FIGURE 6.10

FIGURE 6.11

FIGURE 6.12

FIGURE 6.13

FIGURE 6.14

B. serrata plants usually become fire tolerant by five to seven years of age, re-sprouting after a fire from dormant buds lying hidden under their thick bark (cf. Figure 2.21). As with other trees of this species, *B. serrata* trees are naturally adapted to regular bushfires, with their seed bank being liberated after a fire. In a strong wind, the seeds can be dispersed up to 4 m from the parent tree.

There are more than 700 species of *Eucalyptus*, of which most are native to Australia and many are termed gums (Figure 6.15). They exude large quantities of copious flammable gum from injuries to the trunk, and burning trees can apparently explode. The hollow shell of the veteran *Eucalyptus* tree illustrated here had apparently survived a lightning strike high up on its original main trunk. However, new branches subsequently developed lower down, from dormant buds.

Figure 6.16 shows dead conifers at Opalescent Pool, Yellowstone National Park, Montana. This pool formed in the 1950s, inundating and killing a stand of *Pinus contorta* (Lodgepole Pine) which was growing there. It created this group of white tree skeletons outlined by a rainbow-coloured pool. Subsequently, white silica precipitated on the dead tree trunks and is slowly impregnating and petrifying their wood.

This large branch of *Quercus petraea* (Sessile Oak, Figure 6.17) was broken off from the main tree trunk in a recent gale at Mugdock Country Park, near Glasgow, Scotland. Note the torn open wound revealing elongated strands of the secondary xylem wood; no obvious fungal infection was visible along the broken tree bough.

FIGURE 6.15

FIGURE 6.16

FIGURE 6.17

Figure 6.18 illustrates the rotten, wind-broken hulk of a veteran *Fagus sylvatica* (Beech tree) specimen at Craigmadie water reservoir, near Glasgow, Scotland. This hollow ancient beech tree was infected by several fungal pathogens on the exposed wood of the broken trunk.

A scene of landscape devastation was observed (Figure 6.19) in Stang Forest, northern England, after the clear-felling of a commercial conifer plantation. Note the piles of wood debris remaining after the contractor's bulldozer had "tidied" the site.

Figure 6.20 shows a wind-felled log (dated to the 1600s) of *Quercus petraea* (Sessile Oak) lying in Dalkeith Park Oakwood, Midlothian, Scotland. This Site of Special Scientific Interest has numerous ancient specimens of oak in the parkland between the North and South Esk Rivers. Some of them date back to as early as the 1500s. Their timber was used at the newly built shipyard in Newhaven (Edinburgh) to construct the ship *Great Michael* for King James IV of Scotland, which was launched in 1511.

The specimen of *Q. petraea* (Figure 6.21) has been grown from a Dalkeith Oakwood acorn (cf. Figure 6.20). Its wooden cage protects the young tree from grazing cattle and deer, and it will help to renew the ancient Oakwood.

This tall and solitary *Pinus sylvestris* (Scots Pine, Figure 6.22), growing in Cambridge, England, is almost hidden from view by the close investment of *Hedera helix* (Ivy) around its trunk.

The trunk of this decayed amenity tree (Figure 6.23) had been sawn though at ground level in Cambridge, England. It is colonised by *H. helix*, with the large oval outline of its stump visible at the lower left.

This large epiphytic specimen of *Asplenium nidus* (Bird's-nest Fern, Figure 6.24) is thriving on the branch of a roadside amenity tree at Cairns, Queensland, Australia. Under the moist tropical climate of this region, humus accumulates within the rosette of its basket-like body to provide the necessary nutrients for vigorous growth of the fern.

Figure 6.25 shows the trunk of *Ficus microcarpa* (Strangler Fig), growing in the Daintree Rainforest, Queensland, Australia. Its hollow framework was formed by the roots of an

FIGURE 6.18

FIGURE 6.19

FIGURE 6.20

FIGURE 6.21

FIGURE 6.22

FIGURE 6.23

FIGURE 6.24

FIGURE 6.25

epiphytic fig seedling which germinated in the crown of an adjacent forest tree; the anasto-mosing roots of the strangler then grew down the trunk of the host to soil level. Meanwhile the host tree's vascular system died and eventually rotted away, leaving the frame of the strangler fig in its place.

This trunk of *Pinus nigra* (Corsican Pine), growing in Glasgow Botanic Gardens, Scotland (Figure 6.26), is infected by a monstrous overgrowth. This growth developed from wind-blown spores of the fungus *Taphrina* germinating on the surface of a wound in the Pine's bark. *Taphrina* can attack many tree species (especially birch), but its infection of a Pine is uncommon. The fungal infection spreads relatively slowly and the host rarely dies.

Figure 6.27 illustrates several Mistletoes infesting a farm pastureland tree in New South Wales, Australia. Some 85 mistletoe species have been described in Australia. *Viscum album* (European Mistletoe; Figure 6.28) is an obligate hemi-parasitic plant, commonly infecting a wide number of host tree species in the UK and much of Europe. However, it is rare in Scotland. Mistletoe seeds are spread by birds which eat its white fruits and later void the seeds in their droppings. The sticky seeds adhere to an underlying twig or tree branch. On contact with the host tree bark, the rudimentary Mistletoe radicle slowly penetrates the twig and eventually taps into its vascular system. Pre-Christian cultures apparently regarded the white mistletoe berries as symbols of male fertility, with the seeds resembling semen.

FIGURE 6.26

FIGURE 6.27

FIGURE 6.28

Figure 6.29 shows a dead stump of an *Ulmus* (Elm tree) specimen, in which, under its sloughing bark, numerous fungal rhizomorphs are visible, running along the exposed trunk's wood.

Figure 6.30 shows a large area of soil, now covered by black impermeable plastic sheeting. This was formerly occupied by a grove of ancient Yew Trees, in the grounds of Ross Priory, near Loch Lomond, Scotland. Sadly, these Yews were infected by the soil-borne pathogenic fungus *Phytophthora*. In order to decontaminate the area, the trees were felled and the soil around their roots treated with a commercial fungicide for two years whilst under the plastic sheeting. Only then, could a new grass sward be sown. In the middle distance of this figure, a specimen of the original Yew grove, fortunately free from the fungal infection, is visible.

FIGURE 6.29

FIGURE 6.30

7 Trees and Their Links with Human Cultures

Trees have always had a deep spiritual/religious significance for humans. In Europe, *Quercus* (Oak trees) were important in Germanic and Norse mythology, whilst, in Britain, the Celts believed oak trees to be sacred. Throughout Europe, the evergreen, long-lived, and highly poisonous conifer *Taxus baccata* (Yew tree) was of great significance to our ancestors. Ancient specimens grow in many Churchyards and it seems likely that some churches were established on what were originally pagan religious sites. This Yew, still just surviving, in the Church graveyard at Fortingall in Scotland (Figure 7.1), greatly predates the coming of Christianity to Scotland. In Asia, several species of fig (namely *Ficus benghalensis* and *Ficus religiosa*) are revered and Buddha is believed to have gained enlightenment whilst sitting under the latter species. Such Fig Trees still often have images of Buddha and other religious symbols embedded in their spreading roots, whilst their trunks are bound around with yellow, orange, and red ribbons (Figure 7.2).

Ginkgo biloba (Ginkgo tree, Figure 1.47) still exists as an endangered species in a mountainous area of Zhejiang, China. But it has also survived in Chinese, Korean, and Japanese temple gardens, having been planted there and tended by Buddhist monks. In Chile, *Araucaria araucana* (Monkey Puzzle; Figures 1.23–1.24) is a sacred tree for the Pehuenche Aborigines and its nutritious seeds are a staple of their diet. In the wealthier developed world, recreational activities, such as walking, hunting, riding, and camping amongst trees and forest glades, are important leisure activities (Figure 7.3).

The statue *Salute* (Figure 7.4) was sculpted by Malcolm Robertson in 2007 and is dedicated to the members of the Women's Timber Corps ("Lumber Jills"), who worked hard and unstintingly in woodlands throughout Britain during WW2, and sometimes felled massive trees. This impressive statue stands near the entrance to the Lodge Forest Visitor Centre, near Aberfoyle, Scotland. Figure 7.5 shows the partly trimmed trunk of a veteran *Fagus sylvatica* (Beech tree) specimen still lying in the grounds of Schoenstatt, Campsie Glen, Scotland.

The remains of this ancient several-thousand-year-old male *Taxus baccata* (Fortingall Yew; Figure 7.1) is now enclosed by a iron railings (Figure 7.6). The tree is here viewed from the adjacent Fortingall Churchyard in Perthshire, Scotland. When its details were first recorded in the 18th century, the tree's trunk had split, due to the rotting of its heartwood, into several separate components. However, measurements of the remains at ground level allowed estimation of the original girth of the mother tree to be approximately 16 m; the wooden pegs in the ground demarcate the original boundary of the mother tree.

In Burnham Beeches, Buckinghamshire, England, many specimens of ancient *Fagus sylvatica* (Beech tree) still survive, exemplified by the still-living hollow beech pollard in Figure 7.7.

The stumps of these mature trees of *Populus* sp. (Poplar; Figure 7.8) demarcate trees which were growing at the Te Wairowa village site in North Island, New Zealand. They had grown from Poplar poles which had been planted as fencing by the missionary Seymour

FIGURE 7.1

FIGURE 7.2

FIGURE 7.3

FIGURE 7.4

FIGURE 7.5

FIGURE 7.6

FIGURE 7.7

FIGURE 7.8

Spencer in the mid-19th century. In 1886, a great mud eruption engulfed the village and covered the fencing. The poles (which, by this time, had rooted) were nourished by the mineral-rich mud and, over the next 126 years, many grew into mature trees up to 40 m tall. Sadly, due to decay and wind damage, these trees became dangerous for visitors to the site and so were felled in 2012.

These ancient pollards of *Carpinus betulus* (Common Hornbeam) are still evident (Figures 7.9–7.10) in a former extensively wooded pasture site at Hatfield Forest, in Essex, England. Figure 7.9 shows the trunk of an ancient, partly-hollow pollard whilst Figure 7.10 demonstrates prolific regeneration of a pollarded trunk which had split open. In Figure 7.11, a roadside line of recently pollarded mature *Tilia* (Lime) trees is shown; these were about to sprout in late spring.

FIGURE 7.9

FIGURE 7.10

An internal view (Figure 7.12) of the ancient *Fagus sylvatica* (Beech tree) hedge is shown here. This stretches for over 500 m at Meikleour, Perthshire, Scotland. This "hedge" is composed of over-mature Beech trees, with some reaching 30 m in height. Many of the estate woodsmen were massacred at the Battle of Culloden and maintenance of the newly planted hedge ceased. Today, the trees are trimmed every 10 years, especially on the side facing the adjacent main road. Some specimens have fallen and most of the remainder are decrepit.

Figure 7.13 shows detail of another ancient *F. sylvatica* hedge, growing on a steep roadside bank at Leith Hill, National Trust, England. The hedges were originally planted as protective barriers around new plantations, to guard against grazing deer and cattle. However, decades of long neglect of these hedges followed and subsequent erosion of the bank reveals that many specimens had self-grafted to neighbouring trees.

The ancient remains of a large pollarded *Quercus* (Oak; the Covenanter's Oak, reputed to have been planted in mediaeval times by King David of Scotland) still survives on the Dalziel Estate at Motherwell, near Glasgow (Figure 7.14). During the bloody persecution of Scottish Covenanters in the 17th century, open-air worship under its spreading branches was apparently tolerated. The hollow trunk of an ancient *Quercus* pollard (The Balfron Oak) is illustrated at its site in Balfron, Scotland (Figure 7.15). The trunk, surrounded by

FIGURE 7.11

FIGURE 7.12

FIGURE 7.13

FIGURE 7.14

FIGURE 7.15

the staves of a large oak tub and bound by several iron hoops, apparently served as a temporary prison for local miscreants.

A mature bundle-planted specimen of *Taxus baccata* (Yew Tree; Figure 7.16) is growing at the A82 roadside entrance to a former landed estate at Luss, Scotland. The "tree" is actually composed of several Yew saplings, which were originally planted close together in a ring, but have now thickened and coalesced to form an "instant" mature hollow tree. However, close inspection of the "trunk" reveals a narrow longitudinal gap, where two adjacent specimens had not fully grafted together. At the adjacent estate, several *T. baccata* saplings (Figure 7.17) have been closely planted in a ring; these will eventually grow together to form an "instant" mature Yew Tree (cf. Figure 7.16).

The Romans cultivated olives for many centuries, and Figure 7.18 shows the still-living hollow hulk of an *Olea europaea* (Olive tree) specimen growing in Madeira. Mature, old, and vigorous specimens of *O. europaea* still provide a valuable source of income in the Mediterranean Basin, including Spain (Figure 7.19). These trees have been uprooted from olive plantations in south-eastern Spain. The trees, trimmed of their foliage and larger branches, are now awaiting transport to be replanted elsewhere, to beautify newly-built villas, hotels, and estates frequented by the Spanish tourist trade.

A flowering specimen tree of + *Laburnocytisus adamii* (Adam's tree), growing in the Glasgow Botanic Gardens, Scotland, is shown in Figure 7.20. This small tree (up to 7 m tall)

FIGURE 7.16

FIGURE 7.17

FIGURE 7.18

FIGURE 7.19

FIGURE 7.20

bears three-palmate leaves resembling those of a Laburnum and is an inter-generic graft-hybrid between the two leguminous species *Laburnum anagyroides* and *Chamaecytisus purpureus* (a Broom). It originated in 1825 in Paris from an attempted grafting of the Broom onto the stem of the Laburnum; apparently, the graft failed initially, but later an adventitious bud formed at the graft site. The shoot, which subsequently grew from this bud, showed a mixture of the parental morphological and cytological features. It is now known that plants propagated from the original adventitious shoot (this is only possible by vegetative means) are cytochimeras, with a skin of the Broom surrounding a core of Laburnum tissues. All present-day specimens of the hybrid are derived from the sole original Parisian specimen. Both yellow Laburnum and purple Broom flowers are formed on the tree (Figure 7.21), and sometimes coppery-pink flowers, intermediate between the two parents, also develop.

In Figure 7.22, several Spanish *Quercus suber* (Cork Oak) specimens are shown shortly after their cork was harvested, showing the dark reddish-coloured underlying cork. Cork oaks are medium-sized evergreen trees indigenous to southern Europe and North Africa; they may grow up to 20 m tall and live for several hundred years. In commercial production,

FIGURE 7.21

FIGURE 7.22

cork is first harvested from the tree trunks when they are approximately 25 years old, then the tree's new cork remains untouched until being re-harvested, every 10–12 years, by highly-skilled field workers. They slice through the cork on the main trunk (and often on the branches) and then ease off large sheets of intact cork. Until recently, cork was extensively employed to produce wine bottle bungs; these were cut so that their lenticels ran sideways to prevent gaseous exchange to the outside. However, microbial contamination of the cork sometimes causes the wine to become "corked" and unfit for drinking, hence, the increasing use in recent years of the popular screw-tops for many of the cheaper range of wines. However, steam sterilisation of the harvested cork has recently allowed the resumption of cork use in wine bottling. A close-up view of the trunk of *Q. suber* (Figure 7.23) shows the rough surface of its original bark (top), the newly-exposed reddish bark, which will be harvested in 10–12 years' time (bottom), and the broken face of the newly-harvested cork lying between them.

Making furniture from the pliable wood of *Salix* (Willow tree) is a still-flourishing trade on the island of Madeira (Figure 7.24). Early each year, in the vicinity of Camacha, young Willow branches are harvested from the wild. After sorting by size, canes are stripped of their outer tissues and bundles of them are boiled in troughs of water. The resulting canes are very flexible so that the highly skilled local artisans can more easily weave them into the variety of shapes required for baskets, chairs, tables, and other domestic furniture.

The remains of timber ponds on the southern shore of the River Clyde in Scotland are visible at low tide of the river (Figure 7.25). These ponds extend eastwards for several miles from Port Glasgow along the river's mudflats to near Langbank, Glasgow. The ponds were built in the early 18th century for the seasoning and storage, in the salt water of the River

FIGURE 7.23

FIGURE 7.24

FIGURE 7.25

Clyde, of various species of *Pinus* transported by ship from western Canada and USA. The specially designed ships were fitted with bow doors for loading/unloading the conifer logs, but these ships were highly dangerous for the sailors operating them across the hazardous North Atlantic. The timber was used to construct wooden ships around Port Glasgow, but the trade dwindled after the last wooden ship was built there in 1859.

In Figure 7.26, a recently clear-cut young plantation (only some 35 years old) of *Picea sitchensis* (Sitka Spruce) is shown, and, in Figure 7.27, their stacked harvested trunks are shown. These young trees had been growing near Glasgow, Scotland, but, in their native American habitat, Sitka spruce can survive for 300 years; this fast-growing species may grow by up to 1.5 m per year and reach 90 m in height with trunks 5 m in diameter. Figure 7.28 shows newly planted saplings of *P. sitchensis* growing in the disturbed plantation soil left after harvesting of a "mature" 30- to 40-year-old Scottish tree crop. Sitka seed was introduced into Britain by the Scottish botanist/explorer David Douglas in 1831 and the conifer now thrives in the frequently cold and wet climate of the west of Scotland, where it is grown extensively for timber and for pulp for newspaper production.

A wood-burning kiln (used in the production of charcoal from Pine and other local trees) in the wooded area of Doñana National Park, Spain, is illustrated in Figure 7.29.

Leisure time on a country holiday is often enjoyed in the evenings, sitting around a burning log fire (Figure 7.30). This family fire in the bush, south of Sydney, Australia, was fuelled by *Eucalyptus* logs; note how the heat of the fire opened the wood to reveal several of its radially orientated medullary rays.

FIGURE 7.26

FIGURE 7.27

FIGURE 7.28

FIGURE 7.29

FIGURE 7.30

The foliage on the tips of these tall Palms in Spain have been closely wrapped in bamboo coats to prevent their fronds from greening (Figure 7.31). The etiolated fronds are then cut down and carried, in Catholic religious celebrations on Palm Sunday, to represent the Palms scattered on the ground by the welcoming crowds for Jesus as he entered Jerusalem.

The trunk of this specimen of *Pinus sylvestris* (Scots Pine; the "Faery Wishing Tree"), growing at the summit of Doon Hill, near Aberfoyle, Scotland, is shown in Figure 7.32. This tree reputedly contains the spirit of Rev. Robert Kirk, formerly the Parish Minister in Aberfoyle. In 1691, he had published a fantastic, supernatural book entitled "The Secret Commonwealth of Elves, Fauns, and Fairies", detailing his sightings of and conversations with the fairy population of Doon Hill. When he died, local folklore believed that his spirit persisted within an old Scots Pine located there. Although the current specimen designated as the Faery Tree is still young, its trunk is now richly decorated with colourful messages, wishes and hopes, and various coins have also been hammered into its bark.

In 2015, a weekend Woodland Festival was instituted in the Mugdock Country Park, near Glasgow, Scotland. Hikers were surprised, but pleased, to encounter these colourful additional large "fruits" (Figure 7.33) hanging from several specimens of *Acer pseudoplatanus* (Sycamore tree).

FIGURE 7.31

FIGURE 7.32

FIGURE 7.33

This wooden artwork mask was created by a Costa Rican artist (Figure 7.34) from an obliquely cut trunk of a local rainforest palm tree. "The Gruffalo" (Figure 7.35) was carved by chainsaw in 2012 by Iain Chalmers from the still-standing trunk of a 200-year-old *Fagus sylvatica* (Beech tree) specimen. Sculpting "The Gruffalo" took Iain (helped by an assistant) about four days. It is approximately 5 m tall and stands at the car park entrance to Killmardinny Loch in Bearsden, near Glasgow, Scotland.

This is a night-time view of a tree restaurant in Costa Rica (Figure 7.36), looking through the large branches of a living *Ficus* (Fig tree), at a busy several-floored restaurant located within its heart, showing its happy clientele with their drinks and food.

The prospective balcony terrace to a near-completed multi-storey hotel in Funchal, Madeira is shown in Figure 7.37. Note the construction crane used to swing-up from the ground the numerous, still-small palms in their soil troughs, to the sunny terrace above.

FIGURE 7.34

FIGURE 7.35

A medium-sized maiden tree of *Quercus* (Oak) is shown in Figure 7.38. Several large branches on its left side have been removed by a tree surgeon to allow a power line to safely transmit electricity to the Mugdock Visitor Centre (just visible on the horizon), Milngavie, near Glasgow, Scotland.

A tree surgeon is at work (Figure 7.39), cutting away branches and trimming a mature *Salix* (Willow tree) before the spring flush of new foliage. Note his professional gear (helmet, adjustable visor, saw) and how he has attached fixed ropes from above to catch him in the event of slipping.

A member of a tree surveying team is shown (Figure 7.40), helping to measure the girth at breast height of a veteran *Quercus* (Oak tree), one of several in an ancient tree pasture near Dumfries, Scotland. When, as here, the trunk lies on a slope, the level is taken as halfway between the two levels and the measuring tape around the trunk should be kept horizontal! A rough estimate of a tree's age can be made from this figure, with broad-leaved trees increasing in girth by about 1.5–2.0 cm per year, but, in good situations, such

FIGURE 7.36

FIGURE 7.37

FIGURE 7.38

FIGURE 7.39

FIGURE 7.40

as shown here, the growth rate increases to about 2.5 cm per year. Hence, by measuring the tree's girth and dividing by one of these expansion rates, a good estimate of the tree's age can be made. However, this is only an approximate estimate; foresters can use an 'incre-ment borer' to take a core sample – about the width of a pencil – from a living tree. This is carefully removed and the annual rings in the sample counted under a lens. The hole in the trunk must be plugged to protect the tree against subsequent infection.

Index

T - #0530 - 071024 - C168 - 254/178/8 - PB - 9780367473983 - Gloss Lamination